Tales of the Unexpected

Motorcycle stories with paranormal themes
to appeal to the veteran rider

Murray McLeod

Copyright © Murray McLeod 2018

All rights reserved. This book is copyright protected. Apart from any fair dealings for the purpose of private study, criticism, research or review as permitted under the *Copyright Act (Australia),* no part may be reproduced by any process without written permission from the copyright owner.

This book, and others by Murray McLeod may be purchased on www.amazon.com online bookstores and other retailers.

Photographic acknowledgements
The Motor Cycle magazine
Motorcycling magazine

Other titles
Children's books
Tom, the train spotter

Nonfiction
Aces and Adventurers
Aussie Tennis Greats
Flying Matilda
For Valour
Images of Eagles
Moto Gp
The Unapproachable Norton
TT Legends

Dedication

To all veteran riders and to my old club mates,
staunch companions in a less stressful age

Contents
- Tales of the Unexpected……………………………1
- Spark of Survival…………………………………….5
- High Octane Horror………………………………...38
- The Place that Time forgot………………………46
- Author bio……………………………………….76

Spark of Survival

Mona's Isle; a verdant piece of real estate located in the Irish Sea, with the United Kingdom to its east and Ireland to its west. Since 1907 the Isle of Man was Mecca for the dedicated road racer and despite the absolute beauty of its environment, at times its weather could be described as capricious when rain-laden clouds rolled sullenly over Snaefell, its highest peak. In a single day it could be host to every variable that Mother Nature cared to provide; ranging from brilliant sunshine at sea level to rain-lashed conditions as riders tackled the 2,000 feet ascent of the Mountain. Facing the intrepid racer was thirty-seven miles of the most demanding circuit in the world; with speed variations ranging from 10mph to 120. Included in its challenges were stone walls, trees, telegraph poles and wire fences waiting to entrap the careless rider.

To have even a modest understanding of the course required several visits, yet like some wily female The Island acted as a magnet to ensnare the visitor. Archie Berkley and Harry Lyons were two young men who had succumbed to Mona's spell but for vastly different reasons. This was September, the month of the annual Manx Grand Prix series for amateur riders; the year was 1938 and a time of rising international tensions. The scholarly Berkley came from a privileged background, and on leaving college he enrolled at Cambridge University where somewhat reluctantly he read law. Despite Archie's objections here was a situation that appealed to his solicitor father, for Berkley senior was looking to the day when Archie graduated and became a junior partner in the family practice.

His passion for racing was a situation that his parents were obliged to accept; although it was with great reservation from his father. On the other hand his mother had access to a private income, and actually provided the latest in race ware for Archie to indulge himself in his racing endeavours.

The Young Lions

Harry Lyons could hardly have come from a more diverse background. He was orphaned at fourteen and until completing his secondary education he was cared for by an aunt and uncle. Harry's thoughts were focussed on an engineering career, and on leaving school he was rewarded with an apprenticeship from an electrical engineer. His employer also recognised his enthusiasm for racing and encouraged him to the fullest extent. Slightly built but deceptively strong, Harry was possessed of a natural riding ability and also an innovative nature.

Berkley's attitude was best described as 'languid' and his racing forays were strictly for enjoyment. Occasionally he made an appearance at the hallowed Brooklands speed bowl in Surrey and at short-circuit venues such as Donington Park and Cadwell Park. But for Archie's money nothing matched the sheer joy of batting round the Mountain circuit; where a win in the Manx generally opened the door to a future 'works' ride. That was the pinnacle of Harry's ambitions; however it was a situation he shared with other young hopefuls. Talent scouts from the firms that supported racing hovered in the background as they cast their steely eye on the young lions. In order to make a favourable impression it was necessary for them to ride hard as possible, but without dropping the model or other indiscretions; no team manager likes to see his precious machinery damaged.

By a happy coincidence the unlikely pair booked in to the same Isle of Man boarding house for their initial Manx foray in 1936. The McCalls were a tolerant couple who hosted 'Parkfield', becoming temporary carers for a succession of budding riders, and for a brief period it became their home away from home. No chore was too demanding for Mrs. McCall; even to those 5am practice sessions; she was always there to send them off with that welcome cup of tea. The Archie and Harry duo was an improbable pair but they struck up an enduring friendship despite the limitations of that once a year rendezvous at 'Parkfield'. The innovative Harry was generally at hand to assist Archie in the finer points of machine preparation whenever problems arose. His personal tuning however was conducted in 'cloak and dagger' fashion, away from inquisitive eyes.

No point in revealing secrets to the opposition was Harry's creed.

The Magic Plug

One particular practice session was memorable for Archie, and for a variety of reasons. He was giving the 500 Norton an evening's airing and unlike the previous day when it was going like the clappers, this evening it felt decidedly seedy. Along the short section from Quarter Bridge to Braddan Bridge it hiccupped and misfired like some old crock. He had barely negotiated the left and right kink at Braddan than it spluttered to a halt beside the churchyard.

Archie's mechanical skills were generally limited to the changing of a spark plug; so with a degree of apprehension he whipped out the offending candle and replaced it with a spare from his jacket pocket. Fortuitously that seemed to be problem as he bump-started the Norton back to life. Archie gave a sigh of relief and then settled down to some serious motoring.

He pressed on through Union Mills, Crosby and the full bore section at the 'Highlander' Hotel with its infamous leap. Laurel Bank and Glen Helen received the respect they deserved and then came a brief respite along the flat-out Cronk-y-Voddy straight. For the next ten miles one bend followed another in bewildering succession before the full-bore Sulby straight.

Then it was anchors on for Sulby Bridge; more tricky that day with the late afternoon sun in his eyes. All was well until Milntown when the dreaded misfiring restarted. Archie cursed the situation and for his lack of foresight in having only one spare plug. Shortly the town of Ramsey loomed in his sights; where he spluttered his way through Parliament Square, but the climb up May Hill towards the hairpin was just too much for his refractory steed. He coasted to a stop beside a white-coated marshal.

"Problems chum?" The man enquired. "The roads are opening shortly."

Archie's patience was being tested enough without being reminded that he was stranded on the wrong side of the course. As he leaned the Norton against the kerb he wondered if there was a patron saint for his particular situation; perhaps there was?

He looked up at the sound of a fast approaching competitor and it was Harry; there was no mistaking that distinctive silver and black helmet! Ignoring the marshal's protests, Archie stepped out on the road and vigorously waved Harry down. He slithered to a halt beside his stranded chum, grinning hugely at his misfortune.

"Plug trouble," Archie yelled over the bellow of Harry's exhaust.

"Do you have a spare with you?"

Harry shut down his motor and leant the machine against the kerb. "Sure do," said the quietly spoken Harry.

He delved into a pocket and extracted an unfamiliar item, which looked suspiciously like a chocolate-coated spark plug. Its lower half displayed a threaded metal section while the top appeared to be of Bakelite or something plastic.

"What's this Harry?" Archie demanded. "I don't have time for practical jokes. They are about to open the roads."

Harry made no reply. Instead he swiftly removed the plug from Archie's stricken motor and replaced it with the mystery item.

"Try that," he suggested. "The marshal can give us both a push."

Mystified, Archie replaced his goggles and pulled the Norton back on compression. The uphill gradient made it difficult to gain momentum but when he dropped the clutch he was rewarded by a bellow that only a healthy Norton can give. Still in a state of shock he waited until Harry was also mobile.

Harry pulled up beside him, mouthing the words. 'Let's go!'

The remainder of the lap was a revelation for Archie. Never had his Norton run so willingly; so crisp and clean as he accelerated out of the Gooseneck, through the Cutting and on to the Mountain Mile. Another revelation was to sit in Harry's slipstream and endeavour to stay with him. The sweeping bends past the Stonebreaker's Hut and the Veranda were a challenge. Never before had he essayed them at such a rate and all the time Harry was drawing away.

Suddenly it dawned on Archie that Harry was on his 350 and he was still unable to stay with him on his 500. His appreciation of Harry's riding abilities was being reinforced in a convincing manner; here indeed was a future TT ace! Harry looked back occasionally for a sighting of his chum and obligingly backed off to allow him to catch up. The remaining ten miles were covered at a fast clip rather than a rate that appeared suicidal to Archie.

When they finally stopped at 'Parkfield', two factors were deeply impressed into his thoughts; foremost was the fluidity of Harry's riding; so smooth it was a joy to watch; the other was the performance of his own machine; never before had it delivered such effortless power.

Surely that peculiar spark plug wasn't responsible?

Archie looked up from his musings at Harry's amused expression.

"Enjoy yourself Archie?" he asked wickedly.

Archie gave a sigh and endeavoured to collect his thoughts.

"What on earth have you been up to since last year Harry; don't tell me you're making your own spark plugs?"

Harry gave a confidential smile. "I have in fact. They're not too bad, wouldn't you say?"

Archie was forced to admit they were superior to anything he had come across, but another factor loomed in his thoughts. Did Harry intend marketing his product? If those were his intentions he would need a lawyer to steer him through the minefield of litigation from established manufacturers. Archie felt it pertinent to relay his misgivings to Harry, who was quite aware of the ramifications.

"I simply don't have the resources to market the product. The big boys would crush me anyway."

"Don't be too hasty," Archie suggested. "When I get back home I intend having a word with the Pater. He's always on the lookout to fund a promising scheme. I can handle the corporate side of things; I honestly think we're on a winner."

Archie brushed aside Harry's thanks. "There is another matter old son. Seeing you don't have a machine for Thursday's Senior I'd like you to take over my entry. You'll need to bang in a qualifying lap, but this should be only a formality and I reckon you're in with a fair chance of winning."

"That's very decent of you Archie; my funds never ran to a 500. I'll certainly give it my best shot."

"By the way Harry," Archie enquired. "What brand of spark plug will you nominate at the weigh in? I have an arrangement with KLG."

Harry produced an item from his pocket. "One of these babies; an HLE."

Archie gave a chuckle. "A 'Harry Lyons' plug'; what does the 'E' signify Harry, Eureka perhaps?"

"Excellence, old son. Simple as that, Excellence."

A disappointing outcome

The Manx races are unique in that there is no massed start, with riders being despatched at ten second intervals where it becomes a race against the clock, as your closest rivals may have started ten minutes ahead of or behind you. Consequently it was unlikely that you ever saw them until the race was over. Success in the Isle of Man demanded great concentration and discipline from the rider.

Tuesday's Junior event was blessed with magnificent weather, with barely a cloud in the sky and a pleasant breeze that eliminated the threat of melting tar. For Harry it was a bitter-sweet affair; starting at 18 he rode a copy-book race; forceful enough to put him in third place at the commencement of the final lap, at the same time with due regard for his machinery.

His out of town time-keeper was sited at the end of Sulby straight, and the black board message was brief but positive, '2 minus 10'. He was in second place! A big effort on the last lap and he was there.

It was a day for inspired riding. The earlier fatigue that beset Harry was replaced by that athlete's 'second wind'; a situation more stable than the first adrenalin-charged laps. He swept along the Mountain road, past the Bungalow and tricky downhill Windy Corner, and then faced the exhilarating descent from Kate's Cottage to the slowish right-hander at Creg-ny-Baa, a popular vantage point for a throng of spectators. Programmes, hats and even umbrellas were waved enthusiastically as he swept past, concentrating solely on one target: FINISH!

As he approached the ultra-fast right-hander at Hillberry, Harry made the decision to take it flat out in top gear. In theory it was possible on a 350 but was nearly his undoing as he ran wide on the exit, almost brushing the earth bank with his footrest. With the situation barely under control he was faced with the bumpy uphill left-hander at Cronk-ny-Mona. Somehow that was negotiated, but what an escape! He cursed himself for almost throwing it all away in a moment of indiscretion.

A cold determination replaced the panic from his fright at Hillberry. Signpost Corner with its convenient slip road loomed in his sights. He heeled through in effortless style and prepared for the bumpy and demanding Bedstead Corner. Two more miles and he was home!

The tight approach to Governor's Bridge was negotiated at walking pace; too many races were won and lost at that deceptive right- hander. And all that remained was the uphill run through the Dip, on to the Glencrutchery Road and the chequered flag.

Harry's elation turned to dismay when suddenly the revs. of his willing motor soared. He glanced down to his left, just in time to witness his primary chain lay itself on the road in the style of a lifeless serpent. Robbed of its transmission the Norton coasted to a halt; another twenty seconds and the prize would have been his.

Close to tears with disappointment he began the push to the finish line; a gut-busting uphill effort on weary legs. His frustration was compounded by the passage of other riders; the blare of their exhausts was a further reminder of the fickle nature of his sport. Back at the pits, Mike his pit attendant anxiously watched Harry's scoreboard pointer; stuck fast at Governor's Bridge.

The garbled tones of the announcer were drowned out in the blare of the finishers as they were flagged in. Mike peered along the Glencrutchery Road, more in hope than in the reality of seeing Harry make an appearance, when after what seemed an eternity a lone figure crested the rise and freewheeled across the finish line. Mike dashed across and took charge of the lifeless Norton while Harry made his way to their pit counter. He flopped dejectedly onto the ledge and drowned his sorrows with a bottle of lemonade.

"Hard luck old son," Mike commiserated. "It looked as if it was almost heading for a dead heat for first."

"Really?" he gasped, still regaining his breath.

"You're sure to get a silver replica Harry, although it's miserable compensation for your efforts."

"Let's see what Thursday brings," he said reflectively. "We might be drinking champagne instead of lemonade."

Focus on the Senior

For most riders the transition from a 350 to a 500 represented a quantum leap. It was one thing to wring the neck of the smaller bicycle as you extracted the utmost from it; but far different with the 500 as Harry discovered in the limited practice laps he covered in order to qualify.

Those full bore bends on the 350 assumed far greater challenges on the half litre job, which tended to take charge unless its rider was especially alert. Harry's light physique belied a surprising strength, and was a great asset when forced into a situation that required a certain amount of neck wringing of a wilful mount.

On race eve the riders were obliged to surrender their machines to a large marquee where they remained under guard overnight. Sometimes an extension was granted to a rider should he suffer a last minute mechanical drama. Fortunately for Harry's peace of mind his borrowed mount displayed no tantrums and on race eve he was left to his own resources.

He decided there was no point in opting for an early night when he would only spend his waking hours worrying about tomorrow's race. As a diversion Archie treated him to a night out at the local cinema. Laurel and Hardy were not really their style but at least they provided a laugh or two.

Race day dawned crisp and clear. Harry was awake to greet it.

The forecast of gusting westerlies might well cause problems at exposed sections of the course but at least there was no hint of rain. His race number was 27, placing him in a mid-field situation in a well-supported race. Whatever tension he was feeling on the start line was hidden behind a mask of quiet concentration for the job ahead.

In no time at all he was facing the starter. Down went the flag and with a silent prayer he began a Herculean effort to bump start his mount. There were no problems and he was rewarded with a healthy bellow from an eager machine. The daunting plunge down Bray Hill was his first challenge; with the houses on either side merely a blur of greys and browns as he kept the bucking Norton on line. At Quarter Bridge, warning flags were being waved vigorously as he braked heavily for that deceptive right-hander.

He caught a glimpse of a rider picking up his fallen machine, and then peeled off for the short straight that took him to the left and right kink at Braddan. From there it was it was an exhilarating full bore passage through Crosby and past the 'Highlander' hotel with its heart-stopping leap. It was a day for record breaking and the Norton responded willingly. Already an earlier starter was in his sights and on the approach to Ballacraine Harry made a passing move.

This was nearly his undoing when his late braking manoeuvre almost took him up the slip road. Somehow he negotiated the sharp right- hander with his shoulder practically scraping the hotel wall.

He cursed himself for such a rash move.

Concentrate, you idiot. Don't throw it away on the first lap.

A calculated anger replaced those early adrenalin- charged miles. Ballig Bridge no longer presented a challenge since its hump was eased. The tricky Glen Helen section with its enveloping trees demanded respect despite the dry conditions. There was a brief respite for him on the Cronk-y-Voddy straight and on to more challenges at the Eleventh and Thirteen Milestones.

Mindful of the gusty winds the approach to Kirkmichael village was treated with caution; with the road seemingly narrower than ever in the claustrophobic conditions. Vague images of its residents spectating from their doorways came to him as he swept past at over the ton. Then it was back to open country and never a moment to relax as one bend followed another in an endless procession. His signaller at the end of Sulby straight gave a reassuring wave as he braked heavily for the right- hander at Sulby Bridge; although it was too early in the race for any lap times.

Between Sulby and Ramsey he managed to overtake several earlier starters without further dramas, with each passing manoeuvre giving a boost to his confidence. He swept through Ramsey's Parliament Square and began the climb of May Hill towards the Hairpin. Memories of Archie's rescue mission flooded back. He even received a wave from the same flag marshal who helped them on their way. On the steep approach to Ramsey Hairpin Harry delayed his braking to the very last moment as he prepared to overtake a pair of riders.

Both were travelling in line astern at a modest rate when at the last moment the tail-ender swung sharply to his right. Harry's speed was double that of the other rider and left him with no possible way around the situation. He laid his machine down in speedway style to avoid ramming the other machine. It was a simple enough action as the Norton slid up the road and onto the dirt verge. Harry's elation at retrieving an awkward situation turned sour as he picked up the battered Norton. Such was the impact that the nearside footrest was torn off, but more seriously the brake pedal was embedded in the clutch plates; retirement was the only option. In a bitter frame of mind he wheeled his machine off the course to a safer situation, assisted by a marshal.

"Bad luck chum," he commiserated. "If it's any consolation I'll be reporting that clot to the stewards. Not much compensation for you though."

Harry did not answer. In his heart he was convinced that his last opportunity for a works ride had vanished in that pointless accident.

Different fields

Nineteen thirty-nine represented a year of uncertainty for a generation of young men as Europe headed inexorably towards another conflict. Germany made its intentions clear when it marched unopposed into a helpless Czechoslovakia while Britain and France stood aside. The months leading up to September were periods of rising international tension and were enough to lead to the postponement of the Manx Grand Prix. Harry's disappointment was intense, although for some reason it came as no real surprise.

On September 3 the nation gathered round their radios and heard Prime Minister Chamberlain's solemn announcement that their country was at war with Germany. Almost immediately the British Expeditionary Force was despatched to serve alongside their ally France. Harry could only watch these events from his situation at his employer's electrical engineering works.

His boss had secured a contract for the manufacture of components for the aircraft industry; a situation that placed Harry in a reserved occupation. This was an arrangement that offered opportunities for research and development but Harry was restless for active service.

Archie too was anxious to respond to the call to arms and in his case it was a mere formality. Earlier in the year he accepted a commission in a Territorial mechanised unit, and on the eve of Chamberlain's broadcast he was summoned by telegram to report for duty. Colony Camp was a converted holiday camp set among pine trees in the heart of Surrey, a pleasant spot to begin one's war and being a mechanised unit they fully expected to be despatched to France as fast as transport could be arranged.

Instead, Archie found himself installed as a permanent part of the establishment; with the lofty title of Officer I/c driver training. His Isle of Man exploits had caught the eye of his immediate superior, himself an enthusiast, who was keen to have a TT rider as an instructor.

One of Archie's first assignments was to visit motorcycle dealers in local areas to requisition machinery for army use, and already certain dealers had taken advantage of the purchaser's inexperience in such matters. The outcome found them with machinery that was quite unsuitable for the army's needs. During one of these purchasing forays Archie and his team arrived unannounced at the premises of a South London dealer. As it happened, Alec the proprietor was an ex-TT rider and at once he recognised the lanky Berkely.

"Good morning Mr. Berkley. How can we be of service?"

Archie cut him short. "Don't flannel me Alec. I'm here on a purchasing mission and it seems I'm too late."

Alec nodded in agreement and indicated his near-empty showroom.

"Not much left I'm afraid. Have you tried Comerford's at Croydon?"

"Just came from there," he said grimly. "Anything worthwhile has been snapped up."

Alec made no reply and headed for the workshop area, indicating that Archie should follow suit. He stopped at a flight of rustic stairs leading to another level.

"Alec's Aladdin's Cave," he explained with a chuckle.

Intrigued, Archie followed him obediently up the creaking staircase.

Their trek ended in a walled-up storage area; a veritable bowerbird's delight. From floor to ceiling it was stacked with frames, dismantled engines and gearboxes, even the odd sidecar. Archie shuddered to contemplate the sheer weight of it all on the groaning floorboards. Alec led him further into the maze and finally halted beside an object hidden under a dustsheet. It was so obviously a motor cycle that Archie was becoming impatient with the dramatics.

"I hope this expedition is worth all the subterfuge," he complained.

"Patience lieutenant and all will be revealed."

Alec took hold of the cover and with a Houdini-like flourish laid bare the mystery item. For once Archie was stuck for words. The object of their attention was a rare 500 Sunbeam, looking quite magnificent in its black enamel finish and gold lining. Archie circled the 'Beam in silence, admiring it as one would admire a thoroughbred horse.

"Brand new," Alec assured him. "And one of the last off the line. Do you feel like making me an offer?"

Archie was too engrossed to at first hear the question. He ran a caressing hand over the polished alloy timing cover with its bold 'Sunbeam' lettering.

"What a gem," he said repeatedly.

"A machine for the discerning buyer; someone like yourself," Alec suggested.

Archie looked up blankly at Alec's comment and then suddenly the penny dropped.

He wants me to have it for myself. Colonel Andy will hit the roof!

He considered the options; a lovely Sunbeam and an irate colonel or no Sunbeam at all. There was simply no choice.

"I'll take it Alec. By the way; how much are you asking?"

Retrieving the Sunbeam from its haven proved to be an awkward business and through it all Archie fussed like a nervous parent lest any part suffered the slightest damage. On their way back to camp he made it patently clear to Sergeant Binns that absolutely no one was to lay a hand on his acquisition. Archie could barely wait to start the running-in process.

As he predicted Colonel Anderson displayed a measure of impatience with his purchase, although his ire abated somewhat with the promise of an occasional ride.

That would be after the running-in period naturally!

The Sunbeam and its lanky rider became a common sight on the Surrey byways during that glorious autumn and then all too quickly winter came to northern Europe; the most bitter for forty years. Snow fell everywhere, soldiers in France froze to death in their trenches and parts of the English Channel even iced over. Colonel Anderson chafed impatiently for his unit to be despatched on active service. Daily he badgered higher authority to accede to his demands and just as stubbornly he was refused.

A Wartime Christmas

Those bleak wintry days stretched into wintry weeks and then suddenly it was Christmas Eve; their first wartime Christmas. Colonel Andy was determined to make it memorable for the troops, and thanks to an enterprising QMS there was no shortage of Christmas cheer. An abundance of poultry, ham and plum pudding appeared from out of nowhere to swell the larder.

Christmas day dawned crisp and clear; with the light snow of the evening dispersed by a brisk westerly. A tradition in the armed services was to have the troops served Christmas dinner by their officers, which was also an opportunity for the troops to let off steam and call their officers rude names. Once their duty was completed the prudent thing was for the officers to make an honourable retreat to their own mess. In his role as officer of the day Lieutenant Berkley took no part in these proceedings. Instead he spent the morning confined to the orderly room.

To ease his boredom he was given a mountain of paper work to plough through; a task he detested, while obliging to the end, he was provided with an assistant with a grasp of clerical work. Sounds of revelry from the various messes floated across to the orderly room and its toiling occupants. It was all too much to bear; until finally Corporal McPhee pushed his typewriter aside and moved to the window.

"Lucky sods," he commented in a reference to the revellers. "Let's hope they leave something for us Mr. Berkley."

"I'm sure they will McPhee. Pop the kettle on the stove would you? At least we can enjoy a cup of tea."

"Just a moment sir," he replied. "We're about to have company."

Archie joined McPhee at the window. A lone dispatch rider was heading to their domain, negotiating a wary path on the ice-covered roadway.

What brings a Don R out on Christmas Day? Archie wondered. *It must be damned important.*

The rider slithered to a stop, hauled his machine onto its rear stand and shuffled up the stairs, so rugged up that he resembled the Michelen Tyre Man. Archie looked up disinterestedly at the newcomer, and then to McPhee's amazement he dashed over and impulsively seized the rider in a bear hug.

"Harry Lyons; you old son of a gun! What brings you here?"

The Don R removed his helmet, grinning broadly at his old chum.

"Fiddled my way into a mechanised unit," he explained briefly.

"It's a bit tough sending you out on Christmas Day Harry. Let's hope it was worthwhile."

"Who knows Archie? *Not to reason why*...Isn't that the rule?"

"It does make you wonder at times. I'm certainly tired of hanging around this backwater."

Harry made a quick assessment of Archie's 'backwater', which was not looking the best with its bare trees and snow-covered grounds.

"Make the most of it Archie. When the 'balloon' goes up in France who knows where we'll be after that."

Archie was philosophical about the future. "You're quite right Harry. Those poor sods must be doing it tough 'over there' in this weather."

Reluctantly Harry pulled on his helmet and gloves. "I must go Archie. Two more calls to make before sundown."

He extended a gloved hand. "Take care."

Archie accompanied him to his machine; a rugged 16H Norton, a model that was manufactured in vast numbers for the armed services.

"Still riding a Norton?" Archie observed. "Let's hope we'll both be lining up for the Manx before too long."

Harry merely smiled and prodded the Norton into noisy life.

A brief 'so long' and he was on his way.

Archie waited until he disappeared from sight and then made his way upstairs; and for some reason he was beset by an infinite sadness.

Mediterranean Blues

In no time at all it was March 1940 and with a hint of spring in the air. Those bare trees were already adorning themselves with new leaves. Their 'backwater' was returning to its former beauty, but circumstances were about to change for the wheeled warriors. Suspicions were aroused with an issue of extra winter kit, causing speculation among the troops.

"We'll be off to Norway before you know it," was the general consensus.

"No chance," said other worthies. "I heard somewhere that Russia is about to invade Sweden. That's where we'll be heading."

Only one person came close to the reality. Sergeant Binns was a regular soldier with 12 years' service behind him. Winter kit meant only one thing; it had to be the desert; it was military logic.

Two weeks of hectic activity ensued. Vehicles were re-shod with new tyres, troops endured a series of inoculations to combat strange diseases, incinerators burned late into the night consuming files; as though their past existence was being eliminated.

At last the move for embarkation came and like thieves in the night their convoy moved out of Colony Camp. Never had such a 'backwater' looked more appealing than on that April night and as the vehicles ground past the camp gate, one factor occupied each man's thoughts.

Would we ever see dear old Colony Camp again?

Lieutenant Berkley leant on the rail of his transport, a scruffy freighter of indeterminate vintage and watched the final loading with interest. His precious

Sunbeam was safely installed in the rear of a Bedford 3-tonner. He was finally reconciled to the ignominy of having it painted regulation khaki but it was galling to watch that pristine black and gold finish disappear under the spray gun. Whatever the future held, he was determined to share it with his beloved 'Beam. Around him were sounds of a vessel in port; the grind of winches as their vehicles were hoisted aboard, voices on the darkened wharf below him and squeak of the ship against the fenders. From below decks a chorus of 'Roll out the Barrel' was heard; and then for want of support it trailed away into the night. Archie decided it was time to turn in.

From the confines of his bunk he contemplated his future. He was about to embark on a voyage to an unknown destination. On other vessels men were pondering that same dilemma as they recalled accounts of indiscriminate U-boat attacks in these northern waters. To ease his fears he checked on the proximity of his life jacket, clipped to the bulkhead near his pillow. His cabin mate Jeremy was already asleep and making their space hideous with his snoring. In spite of these distractions Archie also nodded off.

New and exciting sounds invaded his domain. Their cabin trembled with the throb of engines, pulsing through the ship's very soul. Others joined the chorus; a thrum of ventilators and the lap of waters against the quay. He reached out and unclamped the scuttle, in time to see the wharf lamps fade into the distance. The ship began a slow turn, sending a whiff of salt-laden air in his face. They were on their way at last, and in spite of earlier misgivings he fell into the sleep of the contented.

He was awake early and taking care not to disturb Jeremy he slipped on shoes and greatcoat and ventured out on deck. The convoy made a brave sight as it butted into a brisk sou-Easter that created lively whitecaps on the tumbling waves. To his left, a watery sun made a reluctant appearance, silhouetting the ships against dark green seas.

"What a magnificent morning Mr. Berkley." The ship's first officer, Lieutenant Raynor, shattered his idyll. "I trust you slept well?"

"I did in fact," Archie admitted. "But I am rather confused."

"Really?" Raynor replied. "Perhaps I could be of assistance?"

"We seem to be heading due south at the moment," Archie pointed out. "This is completely the opposite direction to Norway."

"Very observant of you, Mr. Berkley; was Scandinavia your destination?"

"More or less," he replied evasively. "At least that was the rumour."

Archie's confusion was a source of enjoyment for Raynor. "Life can be so unpredictable Mr.Berkley. I wouldn't worry too much about it for the moment."

Following breakfast, the entire company was paraded in whatever space was available in the well-deck below the ship's bridge. Colonel Andy took up a position on the foc'sle, armed with a loud hailer.

"At ease men. No doubt you are all wondering about our destination. According to latest intelligence there are developments in the Middle East that could have serious repercussions for the British Empire. Italy has aligned herself with German and is expected to invade Egypt in the foreseeable future. Our Desert Army is well trained but heavily outnumbered by the Italians. An Australian Division has recently arrived and is undergoing further training in Palestine. Our destination is Egypt where we will operate as part of 6th Armoured Brigade with a priority to defend the area west of Suez. I expect every man to acquit himself well in a harsh environment. Good luck to you all."

A great cheer erupted from the assembled troops and it was a mix of relief to know their fate and a determination to prove their worth. That's all very well, thought Archie; but an insidious doubt crept into his thoughts; how would the 'Beam survive, having to ingest that desert sand?

Six days later their convoy arrived at Gibraltar and under the shadow of the 'Rock' the ships were refuelled; with their arrival no doubt relayed to Germany by one of its local agents. To see land again was a great relief, with four-hour leave passes being issued to most of the company; and with them came a salutary warning from their C.O.

Don't get too involved in conversation with the locals. This place is swarming with German spies.

Archie found it refreshing to wander streets that were not blacked-out and to sip the odd Pernod at sidewalk bistros. In the blink of an eye their excursion was over and in varying stages of intoxication they reluctantly filed aboard their transport. Once darkness fell the convoy slipped quietly away with the blue Mediterranean ahead of them, and a day out of Gibraltar their escort of two destroyers was doubled. A pair of 'S' class destroyers appeared over the eastern horizon, cutting a swathe through the sea with their knife-like bows. Heeled over to incredible angles they circled the convoy like zealous guard dogs before taking station on the flanks.

"Proper show-offs that new lot," one soldier commented.

"A life on the ocean wave for you Bert?" his chum suggested.

"Wait until the bombs start dropping," he replied "I'll take mine in a slit trench."

Towards evening they were alerted to the sounds of aircraft engines. An officer followed the newcomer's path with binoculars as it shadowed the convoy at a discreet distance.

"One of ours?" someone asked.

"One of theirs," he answered sharply. "It looks like an Italian Savoia floatplane; keeping tabs on our progress."

With the onset of darkness the aircraft turned away and headed south.

"Making it back to Libya," he commented. "I'd liked to have tickled him up with some ack-ack."

"Bit premature old boy; we're not officially at war with Italy."

"Not yet," he growled. "But it's not far off."

Carefree, sunny days followed with no further alarms; their odyssey had turned into an absolute pleasure cruise.

Here was the lull before the storm.

Sand and More Sand

On May 1st they reached their destination. Alexandria harbour was a ship-spotter's paradise, with the Royal Navy there in strength and bolstered by units of the Royal Australian Navy; a comforting sight for seamen and soldiers alike. Disembarkation took place in the relative cool of the evening, which was a prolonged affair, lasting well into the night, until finally with vehicles fuelled and troops and supplies installed inside, they moved out.

For something like two hours the convoy ground its way nose-to-tail into the darkness, making the affair a trying experience, maintaining contact with meagre, hooded lights in a dusty, stygian blackness.

Finally the lead vehicle stopped in a squeal of brakes. A white-helmeted military policeman emerged from the darkness and moved down the line, issuing terse instructions to each driver.

"This is it lads. Welcome to your new home."

'Home' was a relative term; proving to be a jumble of tented accommodation recently vacated by an Indian unit, and judging by the unsanitary state of the place they must have left in quite a hurry. Days and days of concentrated effort were required to restore it to anything like a civilised situation. In the meantime the engineers set up a mobile workshop section and while not exactly a pristine environment was nevertheless functional, and being a mechanised unit they were often spared the tedium of morning parade. Troops were kept informed with the daily bulletin board and its General Routine Orders; although events involving May 10 1940 certainly warranted a full parade. Looking rather bleak, Colonel Andy addressed them from the rear of a lorry.

News has come through that Germany has invaded neutral Belgium and Holland. The B.E.F. is moving north to support the Belgian army. French troops are engaging the Germans in the south. It looks like the real thing this time.

Uproar broke out in the ranks. Men were clamouring to be where the action was. 'When can we get going?' was the universal appeal.

Anderson raised the hand of authority for silence.

For the time being we remain in Egypt. I have a further announcement to make. We have a new Prime Minister. Neville Chamberlain has resigned and Winston Churchill has been installed as our new leader.

Let's have three cheers for Churchill.

The troop's reaction was overwhelmingly low key; a few ragged cheers here and there but for the most part there was complete silence. Perhaps their fathers were part of Churchill's 1915 Gallipoli adventure; the abortive plan to force a passage through the Dardanelles, which ended in ignominious evacuation. Like their fathers, today's soldiers also remembered the privations of the 1930s under Tory rule. Their lack of enthusiasm at today's announcement was understandable.

Over the ensuing six weeks this was the tenor of the company's existence; a daily announcement of the situation on the Western Front and every day bringing even more depressing news. They were astounded to hear that the B.E.F. was in danger of annihilation at a place called Dunkirk.

Even more amazing was the later revelation that over 300,000 of their number were rescued in Operation 'Dynamo'. In June, France sued for an armistice, leaving Britain to face the Hun alone. On June 10, in a jackal-like move Mussolini declared war on Britain, which came as no surprise to the Desert Army which was more than anxious to inflict a bloody nose on 'Il Duce'. For 18 months Wavell's '30,000' had been preparing for such a showdown, and being outnumbered five to one in no way dismayed them. Their commanders believed in the strategy that offence is the best form of defence and made plans accordingly. But it was the Italians who made the first move when they crossed the Egyptian border and occupied Sollum and Sidi Barrani. British reaction was immediate and decisive; although heavily outnumbered in men and tanks they inflicted punishing defeats in a series of bitter actions.

Here was a demanding environment for a war that was so mobile. If something was essential to your survival you brought it with you. Water, or the shortage of it was of critical concern; petrol too, for tanks and vehicles. Both commodities were an army's life-blood.

In the first week of December, Operation 'Compass' was launched; the precursor of a major British offensive that re-took the territories occupied by Italy in September. An acute embarrassment was the number of prisoners taken; over 130,000 including several generals. In two months Desert Army had advanced 500 miles and was in a position to take Tripoli.

Unlikely Rendezvous

During a pursuit of Italian rearguard elements fate sprang one of its surprises. Archie's situation as liaison officer involved many hours behind the wheel. His preference was the small Commer utility but this day it was a Bedford 3-tonner, a tiring vehicle to drive with its heavy steering and furnace-like conditions in the cabin. The coast road west of Bardia was acceptable in that it was tar-sealed and certainly preferable to the trackless desert they often traversed, although the experience was certainly no holiday excursion. . The mid-day heat rose in waves on the asphalt, creating unbelievable mirages of trees and rippling waters. He crested a rise, shielding his eyes from the glare, and there in the distance was another image, taunting him in its improbability, appearing as some kind of animal, writhing and undulating in the shimmering waves.

He rubbed his eyes, bloodshot and painful from the harsh environment and concentrated even more on the object. Finally it assumed more believable dimensions. His writhing animal was in reality a despatch rider clad in a voluminous coat. His machine stood on its stand minus a back wheel; with its rider in the final stages of repairing a puncture. He looked up from his exertions at Archie's arrival. Recognition was immediate and pleasing.

"Fancy seeing you again Harry! I had no idea they were sending you to the Middle East?"

"Neither did I," he grunted. "I haven't seen my bunk for three days."

Pretty hectic eh; do you need a hand?"

"Not so far but I've lost my pump. Do you have one with you?"

While Harry levered the tyre onto the rim, Archie returned to the vehicle. He was fortunate in his search for it was not uncommon for a tool kit to disappear without trace. It took just a minute or two to inflate the tube and re-fit the wheel. Harry got to his feet and replaced his helmet.

"I'm expected at Sollum right now. Have to go."

He extended a gloved hand.

"Maybe next time I won't be in such a rush."

He booted the Norton into life. A brief grin to his chum and he was on his way. Archie watched reflectively until he was a mere speck in the distance, enveloped in those shimmering waves.

A night of nights

Christmas 1940 was upon them and just as suddenly it was past; so different to the snow-covered 1939 event. The desert war became a Commonwealth affair with the involvement of the Australian 6th Division; and in the course of his duties Archie had established contact with the Aussies. He was impressed with their fighting abilities and amused by their uninhibited behaviour out of the line. For a brief period the combatants adopted a 'live and let live' attitude; perhaps the Christmas Spirit had even reached this harsh environment. The weather too offered a benign atmosphere, in particular the nights. A full moon period was upon them, bathing the desert in cold blue tones, so bright it was possible to read a newspaper with ease. The conditions prompted Archie to pay a visit to his Aussie contemporaries, camped a mile or two from his own. Here was a perfect occasion to wheel out the Sunbeam, and in Archie's estimation the chances of encountering an enemy patrol were minimal. There was even no need for lights, so bright was the moon.

His companions did not share his breezy enthusiasm.

"Watch yourself Archie. The sappers haven't cleared all the minefields yet," was their comment.

Their advice was airily dismissed; according to Archie the night was made for motorcycling. He had traversed the area so often over the past weeks he was convinced he could make the trip blindfolded. Twenty minutes later he was not so sure. Either someone had altered the signposts or he had made a terrible mistake. Finally he stopped in a rock-strewn wadi; facing the reality that he was hopelessly lost.

He became aware of grotesque shapes in his new environment, highlighted in sharp relief by the rising moon. With a start he realised they were vehicles; destroyed in the very minefield into which he had stumbled. A wave of panic seized him. His situation resembled a recurring nightmare of recent weeks and now he was faced with the awful reality. In his confusion he had stopped the engine and in his anxiety to retrace his tracks it refused to restart. Visions of his May Hill experience returned to haunt him but tonight there was no Harry Lyons and his magical spark plug.

Perhaps there was. In the midst of his dilemma he became aware of a familiar figure. It was none other than Harry; parked a short distance away and smiling at Archie's predicament.

He must have crept up on me, thought Archie. *Let's hope he has a spare plug with him.*

He crouched beside the Sunbeam and removed the offending plug; cursing himself for neglecting to bring a spare.

"I've cooked the plug," he called to Harry. "Do you have a spare with you?"

Harry merely smiled and delved into his bulky coat. He extracted a well-remembered object and casually tossed it in his direction.

Archie deftly caught it. He scrutinised the plug; still unable to comprehend how such an innocuous item provided that vital spark.

He felt it pertinent to explain his predicament.

"In case you're wondering what I'm doing stranded in a minefield Harry, I was on my way to 18th brigade H.Q.; do you think you could direct me?"

Harry nodded in assent but remained silent.

Archie turned his attention to the Sunbeam. A quick change of plug and hopefully he could be away from this place. He eased it over compression and then booted it firmly.

Eureka!

She purred like the Sunbeam of old. He looked across to Harry but he was already on the move and heading north out of the wadi.

Take it easy old man! Thought Archie, who was having difficulty keeping him in his sights. They plunged down sandy slopes and along dried-up creek beds at a pace that appeared suicidal to Archie. In his racing days he had never ventured into motocross and now he was being thrown in at the deep end. Somehow he managed to remain upright through the ordeal, to finally slide to a stop beside a group of tents. A group of shadowy figures emerged from the main tent.

"What's the rush Archie," one of them called. "They could hear you all the way to Cairo."

For the moment he was too exhausted to explain. He looked around to thank Harry but to his surprise there was no sign of him. Mystified, he stopped his engine and pulled the 'Beam onto its stand. He removed his goggles and turned to scan the expanse of desert, crystal clear in the moonlight. Not a sound, not a wheel track ….nothing.

"Who were you expecting Archie?" someone asked. "Come inside and join the party."

Still in a state of shock he joined the noisy throng. Someone pressed a drink into his hand. Automatically he downed it, barely aware of its content.

"Have another old boy."

He turned to the speaker; an officer from an armoured regiment and a complete stranger.

"You motorbike types must like living dangerously," said the man.

Archie stared blankly in return.

"You ride full chat through a minefield," the man continued. "Is it some kind of death wish?"

Archie attempted to explain. The best he could manage was a pathetic squeak, but the stranger was waxing loquacious and not to be denied. He waved a hand vaguely in the direction of Archie's route.

"Our Don R was killed out there this morning. Only been with us a week or so; ex-TT rider so they tell me; a chap called Harry Lyons. Did you happen to know him?"

Archie blinked uncomprehendingly at the stranger, feeling as though he had been hit in the solar plexus. Without a word he reeled out of the tent, still clutching his drink. A chill night air caught him in its grasp. He looked up absently at the full moon. So complete was his desolation he might well have been on the moon himself. Sounds of revelry from the tent restored him to a plateau of normality

Something else needed urgent clarification. He crossed to where the Sunbeam stood, cold and inanimate in the moonlight. He knelt beside it, uncertain of what may confront him. Grasping the high-tension lead he followed it from magneto to spark plug. He ran a finger over its contours. No doubt about it, it was a regulation product, the ubiquitous KLG.

Of the Harry Lyons product there was simply no trace.

Bursting with curiosity he switched on the fuel and eased the motor over compression. One gentle prod was sufficient; she ticked over with the regularity of a metronome.

He shut it down, turned off the fuel and turned to gaze into a featureless landscape.

"Thanks Harry," he said softly. "Thanks old chum."

Postscript

Britain may have won its war; surviving the peace became its next priority. Her economy was in tatters, her people dejected. Food was rationed and in short supply, petrol was available only to those in essential situations. Coal miners, wharf labourers and others were imposing newfound powers with constant strikes, to cause additional misery in the heart of a miserable winter.

The Manx Motorcycle Club achieved something of a miracle in overcoming such obstacles; when September 1946 saw the rebirth of the Manx Grand Prix. The machines were basically the same ones that their owners' regretfully put into storage in 1939. Many of the familiar faces were there; a little older but just as keen. There were some missing; those who failed to complete the course.

Mona's Isle was in an intemperate mood for much of the practice period; race days were little better. The Junior event was plagued with patchy rain but for the Senior it pelted down continuously. Early in the practice period a touch of mystery was provided with the appearance of a floral wreath which was first seen against a wall on the May Hill section. Its donor was anonymous and the tribute was brief.

In memory of an old chum. None passed this way more bravely.

A local resident was of the opinion that the donor was a tall, rather stooped man. He appeared to have mobility problems, possibly due to war injuries. Every year a tribute to the unknown rider was laid, and each year it was at a different location. Finally the September pilgrimage was no more. The final tribute was placed on that May Hill section; perhaps the site held particular significance for both of those anonymous souls.

High Octane Horror

The beam of the motorcycle's headlamp played on a car park completely packed with vehicles. Its rider issued a silent curse at the prospect of finding alternate accommodation at this out of the way village; and still the rain pelted down as it had done all day and into the evening.

Ted Gray's visit to the United Kingdom had turned into a generally miserable affair; with a summer that was memorable for its record-breaking rainfall. Many sporting fixtures were in danger of cancellation or postponement; and one event which was threatened was a rally organised for Triumph owners' worldwide. From near and far they converged on Triumph's original birthplace at Meriden in Coventry, refusing to allow the perverse English weather to spoil their day.

Ted made the trip all the way from Australia in a pilgrimage that was far from unique. He found himself as part of a small army of enthusiasts, loyal to the marque; North America was represented, as was South Africa, New Zealand and even Scandinavia. Ted's trusty 1964 Bonneville travelled in the cargo hold of his Qantas Jumbo; this was his first overseas trip and one that involved a great deal of forward planning.

Allowance was made for most contingencies with one exception; the English weather, which had done its best to scuttle his plans. At the rally's conclusion Ted made a hasty decision to head for another appointment at Kettering before darkness overtook him, which proved to be an unfortunate choice when he found himself lost and justifiably irritable after an unpleasant journey.

Ted estimated it was already 11 pm, and closing time at the hotel. He decided to at least enquire about accommodation; even a bunk in the cellar sounded luxurious. After the chill dampness of outdoors the smoky atmosphere of the bar was stifling in its humidity, as he headed towards the counter with helmet and rucksack in hand, and acutely aware he was the object of the crowd's attention.

The mark of the leper, he thought ironically. *One of the ton-up boys who terrify elderly motorists.*

A man clearing a table of its empty glasses smiled a greeting. In spite of his fatigue Ted smiled back and continued on to the counter. Unlike the man, the woman behind the counter was terse and impatient.

"Bar's closed," she reminded him curtly.

"Any chance of a room for the night?" Ted pleaded.

"We're completely booked out," she replied with an air of finality.

That was a disappointing revelation. He collected his gear, and as he prepared to leave Ted found his way barred by the same smiling man.

"Hold on mate," he said in a reassuring tone. "We might have something for you. Just follow me."

Ted followed the man obediently past the counter; receiving a glare from the woman on their way through. They proceeded along a hallway and up a flight of stairs.

The man paused at the top of the stairs and introduced himself.

"I'm Les Harris; Mine Host of this establishment. You look just about all in at the moment."

"I've just ridden through from a rally at Coventry," Ted explained. "I only managed to get myself lost."

"Bad night for motorcyclists mate," Les agreed. "I'm an old rider myself."

That makes a change, Ted thought...*a kindred spirit.*

The infernal room

Their trek concluded at the end of a hallway. Les indicated a set of steps leading to the doorway of a miniscule room, which on inspection proved to be little better than a loft.

"Not exactly Five Star," he suggested in an apologetic tone.

"Any port in a storm Les. I'm grateful to put my head down anywhere."

Ted kept any reservations about the room to himself, finding that it was only possible to stand upright in the area next to the bed due to the steep angle of the ceiling, while its dimensions approximated a prison cell, relieved only by a tiny window that defied his attempts to open it. But its outstanding feature was an all-pervading aroma of petrol and engine oil. Les anticipated his concern and attempted to gloss over the situation.

"We don't use the room because of the smell, and there's no way we can find the source. Do you still feel like dossing here?"

Ted dismissed its shortcomings. "I'll survive. Maybe it was designed for old petrol-heads like ourselves. What's this going to cost me?"

An uncomfortable silence prevailed. Les had forgotten just how grim the room was; prompting him to make a generous offer.

"This one's on the house mate. Hold on and I'll organise a bite to eat. You must be ravenous.

He brushed aside Ted's thanks and made his way downstairs.

Ted took the opportunity to remove his heavy riding boots and waterproof overall. He folded the suit, laid it on the floor beside his helmet and flopped gratefully onto the bed. He would have nodded off only for a discreet knock on his door that heralded Les's return.

"Get yourself around this mate," was the invitation as he laid a tray on the chair beside the bed. Ted was faced with a ham sandwich and glass of milk, and never had a snack tasted so grand.

Les stayed long enough to make a comparison.

"I noticed your Bonnie in the car park. The young barman who bunked here rode a bike too, so they tell me. That was a bit before my time."

"Really?" Ted replied. His fatigue was on the point of overwhelming him. Whoever the previous occupant it was not his concern. For all he cared it could have been Attila the Hun.

Ted's weariness was also apparent to Les. He took his leave.

"G'night then cobber and pleasant dreams."

A night to remember

Ted stripped to his underwear; his preference was to shower and freshen up but fatigue won the argument. He slid gratefully between the sheets, icy cold to the touch, but in spite of his tiredness the events of his memorable day crowded into his thoughts. At first he found it difficult to unwind, until finally he drifted into that soporific state, transported to the fantasy of dream world.

He was at the controls of a powerful motorcycle; not his familiar Triumph but a thumping Norton International. Memories of an earlier era flooded back when he raced such a machine. Agreeable memories were revived as he took it to peak revs through the gears. The empty road flowed like a grey ribbon as he tucked behind the screen, savouring the thrill of a full-blooded racer.

Below his right boot came the blare of the megaphone exhaust; music to the ears as he swung effortlessly through the sweeping bends. Suddenly his view of the road vanished as he approached the crest of a hill.

The Norton cleared it with both wheels off the ground and then plunged down the other side. With mounting alarm he realised how steep was the gradient but more disturbing was the right angle bend at the bottom and a menacing stone wall. Instinctively he sat up and went to close the throttle. A dismay that turned to horror swept over him when it stayed wide open.

He was faced with a rider's worst nightmare as the speed built up at a terrifying rate. His brain searched for options to escape the situation but time was against him. Closer and larger grew the wall until it assumed gargantuan proportions, accentuated by a bellowing motor almost begging for a release from its agony.

Ted had no recall of any impact; only a feeling of absolute terror as he looked about him. He was back in his old room again. But it was a totally different atmosphere and for some reason he felt he was not alone. His physical state felt little better; with his breath coming in gasps, in tune with a heart that was racing fit to burst. Sweat coursed down his face and through every pore of his body, but try as he might he found it impossible to move a muscle.

His body was trapped in an aura that threatened to engulf him.

Waves of nausea swept over him, along with an overpowering force that thrust him into the realms of sleep. He could no longer ignore the challenge. Once again he found himself at the controls of the same Norton, but this time a completely wilful machine. Again they swept along the open road, megaphone blaring but with no control from him; the machine had taken charge completely. A blur of familiar scenery rushed by in a haze of wind and exhaust noise, and then with horror he crested the same hill and began the plunge towards the same unyielding wall.

Once more he was back in the room of horror and unable to escape the recurring nightmare. Sleep claimed him again to face another ride of terror and then back to the straight jacket that was his bed. A voice from his subconscious issued a warning that unless he escaped its environment he would not survive the night.

A drastic remedy

A solution had to be found before his heart succumbed to the bizarre stresses imposed on it. His subconscious drew on a wealth of riding experience to beat the demons at their own game. He lay there waiting, impatient for the dream to restart; to put his theory to the test. It was not long in coming.

Once more he swept along the mountain road at the same unabated speed. He glanced down to his left, to where the magneto lead ran to the spark plug. In spite of it being a dream, his machine was authentic in every detail.

Then with all the strength he could muster he snatched at the lead, ignoring the surge of high voltage. His arm was flung upwards with the blast, but in absolute relief he saw the lead flay wildly in the slipstream.

Deprived of its life-giving sparks his dream machine ceased its taunting bellow and gradually coasted to a stop. The Norton fell to one side while he fell to the other, utterly spent. His final memory was an overpowering smell of racing fuel and hot engine oil.

Ted stirred and took stock of his situation. He was alarmed to find he was on the floor of his old room. His left arm throbbed abominably as if its nerves were in revolt; although it seemed a small price to pay for an escape from his nightmare. He switched on the light and checked his watch, which was showing 3am but he had no intention of returning to his bed, looking a complete shambles with sheets and blankets scattered in all directions. He crossed to the window, frosted up and opaque with condensation. After a few wipes with a towel he was witness to a clear, moonlit morning.

His initial reaction was to dress and put as many miles as possible between himself and his present situation; but his aching arm was a restraining factor. He put it down to the fact that he had slept on it in an awkward manner. All of his discipline was called on to dismiss the nightmare episode.

Ted dressed in a leisurely manner, giving his arm an opportunity to settle down from its ordeal. Surprisingly the overpowering aroma of fuel vapours seemed to have fled the room; he preferred not to dwell on a possible reason.

A tinge of pink appeared in the eastern sky, giving a hint of a lovely summer's day to come.

He made his way quietly to the car park. There stood the Triumph, cold and inanimate and dripping with moisture from last night's rain. He ran a rag over the plug leads and ignition coil to enhance his starting prospects; a prod or two on the kick-start and it burst into life. He took one last look at the window of his room, snicked into gear and trickled out onto the highway.

Quiet as his leaving was, it did not go unnoticed.

Terry the cellarman was up and about, already on the job when his boss joined him.

"That bloke on the Trumpy must have had a better night than the previous guest," Terry commented.

"How's that Terry; it would have been before my time?"

Terry gave a shudder at the memory.

"Couple of years ago it was, and a wet night like yesterday. In the morning the maid found him on the floor, stiff as a board; heart attack according to the ambulance guys. I was surprised when you put that bloke up last night."

"What are you saying Terry; is there a problem with that room?"

Terry was reluctant to expand on its history but Les was not to be denied.

"It goes back to around 1957 when young Smithy came here as a casual barman," Terry explained. "Proud as Punch he was with his Norton International; and then one night he hit this stone wall on the Kettering Road and killed himself. Throttle must have jammed, the Coppers reckoned."

"Don't tell me he bunked in the same room Terry?"

"Only on weekends but the funny thing was, we could never get rid of the smell of petrol after his accident."

Terry looked to Les for a response, but he was already on his way upstairs to close down the cursed room forever.

In the meantime, while Les was putting his domestic demons to rest, Ted was endeavouring to banish his memories of a challenging episode. The traffic-free roads and sunny morning were balm to a motorcyclist's soul, and as the miles effortlessly rolled by he became focussed on more positive thoughts. His experience was merely a nightmare of the most alarming kind, he convinced himself ……..and that throbbing shoulder? That was an obvious legacy of his prang at the old Mount Druitt circuit; so long ago……what else could it possibly be?

The Place that Time Forgot

Under the leafy canopy of a sturdy elm tree the rider surveyed his immediate realm. This was his first experience of a European summer, which in the main had been pleasantly mild and rain-free. However the approaching autumn was punctuated by periods of sudden showers; welcomed possibly by garden gurus but certainly not appreciated by active motorcyclists. His present situation was a result of such a downpour, prompting him to take cover under a convenient tree until this local shower abated; besides he was forced to admit he was sorely in need of directions to find a somewhat elusive character in this out of the way south coast village.

Rob Barrett's affiliation to the two-wheeled brigade was on a professional basis; as a motorcycling journalist, which involved reporting on motor sport and road-testing new models coming onto the market. This situation, lofty as it may imply had only been achieved by an unswerving determination on his part; and after a hectic European schedule his mission today was a final pilgrimage before an eagerly anticipated return to Australia. Meantime, the rain showed no sign of abating, prompting him to settle down for the moment under a waterproof cover and hopefully catch up on some much-needed sleep. Drifting off into that soporific state he was gently transported into an earlier era, of a young man, impatient to become a member of that 'fourth estate'.

Being an only child was no impediment for Rob Barrett, it was in fact a stimulus to his outlook without the presence of any peers. Another factor was the absence of a father figure in his formative years. Barrett senior was certainly no role model in the youngster's upbringing, in particular those early teenage years when a father's influence should have been most needed.

Modest though it was, their inner-city semi-detached cottage was at least a tangible asset for Rob and his mother Aileen when Alf Barrett made the decision to vacate the family home and move permanently to sunny Queensland. The promise to 'love, honour and obey' in his wedding vows was conveniently dismissed as their once-blissful marriage lurched from one crisis to the next. For whatever reason; Alf was irresistibly attracted to 'those other women', who in turn returned their favours quite freely. Finally Aileen called 'enough' to these dalliances, and despite assurances from Alf that he was a reformed character, his latest liaison was the last straw for the tolerant Aileen, who felt no pangs in consigning Alf over the border to Queensland or anywhere as far as she was concerned.

For his part, Rob's attitude was more of disappointment than anything else in the outcome, and certainly strengthened the bond between mother and son. At the time Rob was aged fifteen and studying hard for his Intermediate Certificate and the last distraction he needed was this latest one. In fact, being called 'Rob' was simply a compilation of his given names; Roderick Oliver Barrett, which transposed quite comfortably to 'Rob', a name he was quite happy to carry throughout his early years and into a journalistic career.

At that impressionable period of Rob's teenage years, three people, all of vastly different outlook were to influence his future career. Foremost was Aileen, whose unfailing support kept him on an unwavering path throughout those years of turbulent domestic travail. Secondly he had good cause to thank his high school English master, the unsmiling but caring Mr. Lamington. Most school masters acquire nicknames, in many cases unflattering ones, while in Mr. Lamington's situation it was inevitable that he was labelled 'Cakey'. The humble 'Lamington' features large in Australian folk lore, this gourmet's delight being derived from sponge cake cut into squares and coated with chocolate and shredded coconut. Any self-respecting Aussie afternoon tea was deemed incomplete without the ubiquitous lamington.

Scholastically, Mr. Lamington was quietly impressed with young Rob's literary talents, in particular his lively essays. In fact it was 'Cakey' who suggested that Rob should pursue a literary career, or alternately, move on to university to become a lecturer or school teacher. This last proposal was tempting, however this involved going on to gain his Leaving Certificate, plus considerable financial stress until that university degree was achieved. The third person involved in Rob's development was an unlikely character, an irreverent type, not highly educated but well able to cope with the challenges that life threw at you from time to time. He was in fact Aileen's older brother, Charlie Edwards, a largely self-taught motor mechanic and general knock-about handyman. Charlie never married, being quite content to live in a flat behind his modest service station and workshop located just out of Bathurst in the central west of New South Wales.

On occasions Rob would spend his school holidays at Uncle Charlie's establishment and despite the rather Spartan accommodation, the experience could be quite uplifting. For Rob to actually reach Charlie's garage involved a four hour coach trip from Sydney, and with the garage located on the Great Western Highway the commuter was delivered virtually to the doorstep. Although somewhat in awe of Charlie, with his fairly unconventional life style, Rob managed to adapt to these new surroundings; to the extent of becoming involved in petrol pump duties and lending a hand in the workshop. This atmosphere was agreeable enough for Rob to reconsider a career in mechanics, as opposed to his journalistic aspirations. He reasoned that a competent mechanic would always be in demand, while conversely a journalistic career could never promise an iron-clad security.

Another attraction of Charlie's establishment was his collection of vintage motorcycles, amounting to ten or so in varying stages of mobility and housed in a fair-sized shed at the rear of the property. Some were actual runners, while the majority were sorely in need of mechanical ministration to make them mobile again.

However, one machine stood out from the pack; a classic International Norton, pristine and gleaming in its silver and black livery and distinctive red pin striping to the petrol and oil tanks. Here was Charlie's obvious pride and joy, which he was always ready to impress on Rob whenever they spent time in the hallowed grounds of Charlie's museum in miniature.

"I call her 'Clattering Kate'," he explained to Rob. "Bought her brand new in 1937, which now makes her exactly ten years old."

As Charlie removed the dust cover, Rob stood in admiration, regarding the Norton as one would admire a fine horse. "She's a beauty alright," Rob agreed. "When do you find the time to ride her, Uncle Charlie?"

"Whenever I get the chance Rob," Charlie replied."By the way, how do you fancy having a go around the paddock on the old M20?"

The machine in question was an ex-army BSA, still in wartime khaki but nevertheless a runner, and despite its bulk was docile enough for a beginner to make his first essay on two wheels. Before Rob had time to consider the proposal Charlie booted the BSA into noisy life and sat Rob at the controls. The cockpit drill was brief and to the point and before he realised it Rob was under way, with an encouraging farewell from Charlie.

"Off you go then and try not to bend it."

Meanwhile Rob had managed to get under way without stalling the engine and as he negotiated the bumpy paddock was enjoying the moment. What a thrill as the urge was wound on and the scenery flashed by, even though he was still in first gear, wary of changing into second! After ten minutes or so of paddock bliss, somewhat reluctantly he returned to his departure point.

"Enjoy yourself son?" Charlie enquired.

Rob did not respond; his ear- to- ear grin said it all.

For the remainder of the school holidays the BSA became the focus of Rob's endeavours, each paddock jaunt adding to his confidence on two wheels. Then all too soon it was time to return home and face the reality of study and a career choice.

The Heritage

Enjoyable as it was, Rob's holiday sojourn had the effect of unsettling him in his scholastic aspirations. Those blissful paddock outings on the venerable M20 had ignited an abiding interest in things mechanical, while at the same time his journalistic aims still burned strongly as ever. Now in the third term of Year 9 there was the approach of that all-important Intermediate Certificate, where a satisfactory pass could open the door to a variety of career options, however a successful outcome in the Year 12 Leaving Certificate offered far greater ideals.

With Alf Barrett a *father in absentia*, his English master, the redoubtable Mr. Lamington had assumed the role of father figure in Rob's domain, and to whom he turned to for career advice. Unsurprisingly 'Cakey' expressed a degree of disappointment at Rob's suggestion that he was considering leaving school following the Intermediate Certificate. On reflection he came up with an interesting compromise, whereby Rob could pursue a journalistic career, and as his experience expanded he might possibly make a specialty of reviewing events in the motoring world.

This new option sounded most appealing, although it all depended on gaining a position with a news outlet and from there eventually moving on to greener pastures. In the midst of all this decision-making a significant scenario arose, which involved Uncle Charlie and one that would create further diversions around Rob's future.

Returning home following a mid-week day of unrelenting cramming for that Intermediate Certificate Rob was disconcerted to find Aileen seated at the kitchen table in a tearful state. This was an unusual state of affairs for the generally optimistic Aileen, causing Rob to fear the very worst. Drawing up a chair beside her Rob took her hand.

"What's up Mum?" he asked gently. "Have you had bad news about Dad?"

Aileen smiled faintly at the suggestion.

"Nothing like that son, it's your Uncle Charlie. I just received a telegram that he passed away suddenly two days ago at his workshop."

"Uncle Charlie?" Rob echoed. "I always thought he was bullet-proof. Who was it sent you the telegram?"

"You remember Harry his offsider? It appears he found Charlie early the other morning on the floor of his workshop, they say it was a massive heart attack."

Last will and testament

Until Rob's recent holiday break, his Uncle Charlie had been something of a remote figure in the scheme of things and with his death Rob's initial reaction was to experience sympathy for his mother. A secondary emotion was the fate of that motorcycle collection, which hopefully would not be consigned to the local tip following Charlie's funeral.

As it happened, that event turned out to be a non-event for Aileen and Rob, with Charlie's solicitor notifying them that there would be no funeral service. Instead Charlie had suggested that the family might care to conduct a private wake and at their leisure.

That revelation was somewhat sobering to the family; however there was still unfinished business, with a notification that they attend the offices of Densley and Densley for the reading of the will.

Mr. Alec Densley displayed something of an intimidating image as they presented themselves at his city premises. Tall and sombre and immaculately attired in a three-piece business suit, Rob was immediately reminded of his English master, the unsmiling 'Cakey' Lamington. First impressions are not always lasting ones, for Mr. Densley exuded kindness itself as he outlined the details of the estate of the late Mr. Charles Edwards.

"The service station and premises located on the Great Western Highway at Bathurst," Mr. Densley began. "I bequeath to my business partner, Mr. Harold Alexander."

Densley looked up to study their reaction and then continued.

"Bank deposits to the value of one thousand five hundred pounds are to be made available to my sister Aileen Barrett."

Aileen's heart virtually missed a beat at this disclosure; remembering Alf's departure which left her to maintain an outstanding mortgage on their modest semi-detached cottage.

"A most welcome piece of news Mrs. Barrett?" Densley suggested.

"I recently received a last-minute addition to this document," Mr. Densley continued. "And one that concerns you Mr. Barrett."

"Mr. Barrett!" Rob felt thoroughly uplifted by the title, but what on earth could this mean? He was not left long to ponder the issue.

"It appears that Mr. Edwards was the owner of quite a large selection of motorcycles and Mr. Edwards is offering three pieces of that collection to his nephew, Roderick Barrett."

It became Rob's turn to experience heart palpitations.

"These items are listed as a 1937 Norton International, a 1942 BSA M20 and a 1914 Norton Brooklands Special," they were informed.

"This is an unusual bequest", Mr. Densley suggested. "Are you in a position to accept these items Mr. Barrett? I imagine you would be too young at the moment to hold a motorcycle rider's licence?"

"I'm still in a state of shock Mr. Densley and no I don't have a licence, but how can I access these items?"

"Mr. Alexander is prepared to have then transported to your home address whenever this is convenient. Shall I notify him to proceed?"

Rob could barely hide his glee at this windfall, Aileen on the other hand was decidedly uneasy at the prospect of three ancient motorcycles being landed on their doorstep.

New Fields

The eventual arrival of Charlie's bikes was an affair never to be forgotten, and through it all Rob fussed anxiously as they were unloaded from their transport. Space for the precious Inter was found in the backyard shed, while for the moment the BSA and Brooklands Special were installed under the landing of the back veranda. Hopefully this would only be a temporary arrangement and at least they were not exposed to the elements.

Meantime the Intermediate Certificate was conducted during mid-October and until the results were published in late November, the participants were obliged to bide their time in varying moods of optimism or pessimism. Finally that day arrived, one that saw Rob and Aileen anxiously scanning their copy of the Sydney Morning Herald. For some students it was a moment of triumph, while for others a veritable mish-mash of results, the proverbial 'curates egg'; good in parts. In Rob's situation it was more than satisfactory, four 'A' passes, including the all-important English paper, while the 'B' pass in Latin was no great disappointment, that English paper was the apex of his endeavours.

Some days following the Intermediate results a letter arrived, addressed to Mr. R.O. Barrett, and at once Rob recognised that distinctive writing style of R.H. Lamington *M.A.* Bursting with curiosity he opened the envelope, quietly touched that his old English master had taken the time to contact him. A further mystery was provided by a second envelope included with the original and addressed to a Mr. Charles Markham.

Mr. Lamington's letter offered optimism for the tyro journalist.

My Dear Roderick,

First let me offer my congratulations on a pleasing Intermediate result. As you are aware my preference would be to see you continue your High School studies and gain that Leaving Certificate, however I can appreciate your ambition to launch yourself on a journalistic career. Perhaps I may be able to assist you in taking that first step. Attached is a letter of introduction to an associate of mine from those far-off University days. Charles is managing editor of a newly-instituted daily newspaper, 'The Evening Mirror' and I recommend that you make contact ASAP and present that letter. In the meantime I wish you all the very best in your endeavours.

Sincerely

R.H. Lamington

The Mirror beckons

Situated on Broadway, conveniently near Central Railway Station, the 'Evening Mirror' was well-placed for an inner-city news outlet. To the rear of its imposing facade, a large warehouse complex had undergone a major update to create the production side of its operation, while an expansive yard area housed the fleet of delivery vehicles, resplendent in their Royal Blue and Silver livery.

As Rob waited in the reception area for his 10am appointment with the managing editor, he was still coming to terms with the sheer pace of recent events, while still harbouring niggling doubts of ever achieving that journalistic dream. His pessimistic musings were interrupted by the arrival of the Great Man's secretary.

"Mr. Markham will see you now Mr. Barrett."

Solidly built and prematurely balding, Markham displayed a prosperous mien as Rob was ushered into his office, which overlooked the news room, with its reporters beavering away at their typewriters. Despite the early hour the atmosphere was suffused with cigarette smoke, punctuated by the constant ringing of telephones and noisy banter of conversation. His office was shielded to a degree by double-glazed windows, which made the situation at least tolerable.

Rob was somewhat surprised to see a manila folder on the desk, with the title 'R.O. Barrett' boldly inscribed.

"I've been perusing some of your high school essays," Markham commented. "Our mutual friend Mr. Lamington was kind enough to forward them recently."

"And how did you like them?" Rob enquired, wondering how he had summoned the nerve to ask such a question.

Markham permitted himself a wry smile. "Very much, and Mr. Lamington is of the opinion that there could be an opportunity for you here at the 'Mirror'."

"I do have a letter for you from Mr. Lamington," Rob ventured. "Would you care to read it?"

Whatever was contained in that introductory letter, for Rob it was a career-opening option.

"Good old 'Cakey'" Markham chuckled as he swiftly scanned its contents. "This is high praise indeed."

Eyeing Rob steadily, Markham remained silent for what seemed an eternity and then rising to his feet he offered his hand in a warm handshake.

"On behalf of my paper I'm prepared to offer you a position as cub reporter attached to the newsroom. Are you happy to accept this situation Mr. Barrett?"

"*Cub reporter!*" thought Rob. "*What an opportunity!*"

"I'm delighted to accept the offer Mr. Markham, and would it be possible to make a start today?"

"I admire your zeal young Barrett," Markham smilingly replied. "Let me introduce you to the team, and a word of advice Rob," he continued."Reporters have a reputation for letting their hair down in off-duty moments and giving the amber fluid a nudge, so try and maintain an abstemious outlook in the future."

Markham's Minions

The arrival of a cub reporter into the ranks of seasoned newsmen elicited a general attitude of 'who cares' among those worthies. They were for the most part a group of hard-bitten characters, mainly cynical of the current state of affairs in a post-war Australia. One of them seemed to defy that attitude; his features generally adorned with an almost permanent smile, and when conditions warranted light relief, Taffy Griffith was there to provide a song or two.

The irony of the situation was that Taffy's brief was to maintain contact with police and local hospitals to report on the latest crime and accident incidents, and Taffy's presence with his beaming smile would be seen to be incongruous at a grisly murder site. Nevertheless, Rob was initially assigned as his side-kick, which involved their attendance at major incidents, usually at inconvenient times of day or night.

The 'Mirror's' other reporters were specialists in their particular field, typified by Peter Wilson in his role as foreign affairs and political correspondent. Frank Hyde's brief was the local sporting scene, covering rugby league in the winter season, cricket during summer and tennis coverage when the Davis Cup featured strongly in the Australian psyche. Stock market and financial issues were handled by Charles Markham, whose editorials were directed firmly against the policies of the incumbent socialist Labor Party. Some of the 'Mirror's' funding was provided in clandestine donations from the opposition Liberal party, and Charles was looking to the day when they would be swept back into power.

Local fashion and social events were covered by Tania Verstak, a recent migrant from Eastern Europe, and her comely presence added certain freshness within a generally all-male domain. Her employment was on a casual basis, and she found the time to sharpen Rob's prowess on the typewriter in after-hours tuition, even instilling a grasp of the mysteries of shorthand as well.

The Open Road beckons

On the home front, Rob's priorities were focussed on two items, and having turned sixteen he was keen to graduate to a motorcycle licence, although Aileen's enthusiasm did not match Rob's in this venture. Secondly, to achieve that goal he would need to hone whatever skills he acquired on those paddock jaunts aboard the khaki M20. Before that took place, there were shortcomings related to the M20 that demanded immediate attention. His first hurdle was the fact that it was currently unregistered and to achieve that status, an amount of TLC was called for.

Rob was fortunate in that his next door neighbour had been a keen pre-war motorcyclist and now in active retirement Syd. welcomed the opportunity to return the BSA to a new lease of life. A mechanical strip-down revealed no nasty surprises, and apart from a de-coke and valve grind, very little mechanical work was required. Following an oil change, a wash and polish and fitted with new tyres the veteran passed its registration test with no dramas.

Having a street-legal set of wheels was the impetus Rob needed to launch him on the path to gaining his licence and fortunately for his peace of mind during his learning period road traffic had none of the density of later years. The local police station handled licence testing, which involved an oral test, after which the applicant carried out some road manoeuvres under the steely eye of the law. In Rob's situation his capabilities seemed to satisfy the authorities and with a sense of satisfaction he could now sample the freedom of the open road.

An independent spirit

Meantime Rob's career at the Mirror had advanced to the stage where he was entrusted with his own weekly column. This had come about following his proposal to the editor that a series of human interest stories centred around those of the community who had currently fallen on hard times could be a stimulating feature.

Markham agreed that the proposal had a future and as a result Rob's energies were directed to conducting interviews at evocative places, such as the Children's Hospital at Camperdown, the War Veteran's Hospital at Concord and others. Such was the public response to his column, 'Rob's Ramblings' that it became a daily rather than a weekly feature. To add reality to the column he had invested his hard-earned wages in an impressive 35mm Zeiss camera to add a visual aspect to articles redolent with both pathos and optimism.

At the same time his thoughts were focussed on the prospect of including Australian motor sport activities on the Mirror's sports pages, possibly as a weekend feature. Markham however saw little value in the scheme, urging Rob to concentrate on those human interest stories that were enjoying general public acclaim. Rob, for his part had opposing thoughts, aware of a recently-launched monthly publication, 'Motor Sport in Australia'.

After much personal consideration, Rob made contact with the magazine's editor, whose office was located at North Sydney, which involved a short ride on the faithful M20 across the Harbour Bridge. Compared to the 'Mirror's' plush facade, 'Motor Sport's' office presented a 'country cousin' type of atmosphere; however there were no doubts concerning their enthusiasm for the project. The editor, Doug. Blair, with his black horn-rimmed spectacles, crew cut and turtle-neck sweater reminded Rob of the all-American college graduate, and it appeared that Rob was already known to him.

"I've been following your articles in the Mirror," Blair commented, adding facetiously. "And did you have plans to incorporate them in our glossy mag.?"

Rob laughed at the suggestion. "Not at all Doug; what I'm keen about is to find an outlet for motorcycling articles, you know, road tests, personalities and racing, that sort of spread."

Blair pondered for a moment on the prospect. "Someone in Melbourne is already covering that field," he observed. "Do you imagine there could be enough local interest to support another one?"

"Absolutely!" Rob declared. "And I'm aware of the Victorian one, which is really only a tabloid newspaper."

Blair agreed with Rob's comparison and to his surprise came up with a positive offer. "Can you provide one or two relevant articles with pics. etc. before the end of this month? If my partners are agreeable we could make this a regular feature, and as a rule we pay £20 per article."

On his way back to the office Rob pondered on how best to inform Charles of this new development, although as it happened he need not to have felt apprehensive about his reaction.

"Good for you Rob, syndication is the way to go these days, and if you're agreeable you might prefer to write for the Mirror on a freelance basis."

Free as the breeze

To launch his connection to Blair's magazine Rob hit on the novel idea to conduct a road test using his classic pre-war Norton International. In the interim, neighbour Syd had carried out a comprehensive check-up of 'Clattering Kate', and fitted with trade plates Rob made his way to outlying suburbs on Sydney's western fringe to carry out handling and fuel consumption tests. Syd had followed the procedure from the cab of his Bedford utility, riding 'shotgun' and handling the photo shoots. For the maximum speed tests the abandoned wartime airstrip at Castlereagh was utilised, which was a new experience for Rob to record speeds approaching the magic 'ton' on the historic bike. After several full-bore runs, an average speed of 94 mph was determined, and feeling dual emotions of exhilaration and utter fatigue he opted to load the Norton in the back of the ute and return home in relative comfort.

Feedback from Rob's articles was generally positive, and emboldened by these results he made the decision to conduct speed trials with the venerable Brooklands Special, 'Bessie'. Syd was at hand to prepare the bike for the occasion and pointed out any shortcomings which could arise. With a past stretching back to 1914 the machine's robustness to handle speed trials was not above suspicion, while the spidery bike lacked refinements such as lights, mudguards or front brake. A pitiful bicycle-style rear brake was fitted and its transmission was via belt drive to the rear wheel, with single-speed range.

When new the BS was sold with a certificate guaranteeing a speed of 60mph but in reality there was really one purpose for this old dinosaur and that was on the broad expanses of the now defunct Brooklands speed bowl, hence the title 'Brooklands Special'. Syd's main concern was the ancient beaded tyres, for which there was no current replacement, prompting Syd to insist that Rob invest in a set of leathers and crash helmet.

Castlereagh airstrip was deserted for their mid-week essay, and after donning his riding gear Rob carried out some exploratory runs to come to terms with 'Bessie'. Syd was kept fully occupied with camera and stop watch, while for Rob his test runs were proving far from enjoyable. With its basic suspension, lack of brakes, self-opiniated steering and straight-through exhaust the speed runs were uncomfortable in the extreme.

Having just completed his fourth timed run, the sorely-stressed engine finally cried enough, with an ominous silence replacing the blare of the open exhaust. Rob coasted to a stop beside a concerned Syd, who gave a rapid assessment of 'Bessie's' condition.

"It's the magneto," he declared. "The outer casing has disintegrated and there's no way of repairing it. Unless we find a replacement, I'm afraid 'Bessie's' glory days are over."

All things considered, the BS speed trials had ended in rather unfortunate circumstances, causing Rob to reconsider submitting the article. Blair however was quite enthusiastic about the project; that such an historic machine could be resurrected for its brief moment of fame.

The ramifications of these latest events proved to be far-reaching, all the way to the United Kingdom in fact. Rob first became aware of their significance on receiving a phone call from an excited Doug Blair.

"I've just got a telegram from an English publishing house, Temple Press; you know, they put out 'Motorcycling and 'The Autocar'. You had best get over here to our office ASAP."

Sure enough, the telegram was from Temple Press and 'Motorcycling's' editor was keen to meet the author of those latest articles. A telephone number was provided, with a recommendation as to making contact at a time mutually agreeable to both parties.

"How about that Rob?" Blair gleefully exclaimed. "You're moving into the big time now. Let's make that call to London this evening."

Despite the tyranny of distance the phone call was reasonably static-free. Graham Goodman the editor was able to relay to an astounded Rob that he was invited to be the guest of Temple Press for the period covering the Tourist Trophy races, plus one or two continental GPs. Personal transport would be provided and overnight accommodation at reputable venues. In return Mr. Barrett would grant Temple Press exclusive British rights to his articles, while travel expenses to and from the United Kingdom would be borne by Mr. Barrett, and also his attendance in London would be required no later than June 1. Would these terms be acceptable?"

In a voice not quite his own Rob stammered out a grateful affirmative reply, at the same time rapidly computing his immediate priorities to make that step.

So this is England

"Good morning ladies and gentlemen, this is your captain speaking."

The measured tone of their pilot had the effect of startling a somnolent Barrett into a state of pulse-racing awareness. His immediate neighbour favoured Rob with a pacifying grin.

"That gave you a fright mate," the man observed.

Embarrassed, Rob responded with a wry smile and consulted his watch which read 6am.

"We'll be landing at Heathrow in thirty minutes," the man informed him. "I reckon you missed that last bit."

"We're here!" Rob thought, almost in disbelief. He gazed out of his window situation at those supple wings that had borne him on his odyssey, as they gently flexed in the morning thermals, highlighted by the rays of the rising sun.

"Was it only a day and a half?" he recalled, that he embarked on his inaugural flight?

His magic carpet was a Qantas Constellation, waiting patiently on its tricycle undercarriage at a bustling Mascot terminal. In the weeks prior to this, life had been hectic in the extreme, what with passport and visas to be arranged, plus the dreaded inoculations, which invariably left the recipient confined to bed for a day or so. And now as their 'Connie' touched down at Heathrow, Rob was facing his immediate future with anticipation.

Negotiating customs was a tedious enough business, with baggage checks and the like, however there were no dramas and on leaving customs Rob was pleasantly surprised to be greeted by a staff member from Temple Press. In no time at all a taxi delivered them to an address in Bowling Green Lane, to the hallowed portals of Temple Press.

On meeting editor Graham Goodman Rob was reminded of his initial interview with Charles Markham, a day of much anticipation, which could also be applied to this new scenario. Goodman greeted Rob affably enough.

"Welcome to the UK Rob, and I hope you're not too weary from your flight."

"I managed to grab some shut-eye at the end of it," he informed Goodman.

"That's good," he replied. "Because you'll be heading for Liverpool on the mid-day express train today, for a rendezvous with the Isle of Man ferry."

"On my own Mr. Goodman?" Rob ventured.

"You won't be lonely Rob," Goodman assured him. "We have a compartment booked for you and four other reporters covering TT week. I'd like to be going myself."

In his pre-teen years Rob was fascinated by photos and art works of those English express steam trains, with their evocative names, 'Flying Scotsman', 'Royal Scot' and others, as they thundered their way to northern destinations. Privately he dreamed of the day when he might be part of that experience, and now it was actually happening, but with one major difference. Diesel/electric had largely replaced steam as the prime mover and rather than pound majestically along those rails, nowadays they virtually whirred their way along, and at impressive speeds too. From his window seat he watched an unwinding vista of verdant fields and neat hamlets and cottages, so different from the Australian landscape, vast and arid by comparison.

His travelling companions apparently did not share Rob's enthusiasm for things rural, preferring to concentrate instead on a boisterous round of poker. The rapidly passing scenery began to have a stroboscopic effect on the visitor, and coupled with an on-going fatigue he could no longer fight it off. His last memory was a comment from one of his journalist contemporaries.

"It looks like our young wallaby is about to bite the dust."

"Rise and shine digger, there's a ferry out there waiting for us."

The speaker was shaking Rob none too gently. Reluctantly he opened his eyes, confused at first with his current environment. He looked up owlishly at the man. In this instance it was Bill Hanks, the official photographer.

"I'm sorry Bill, my body is all out of sync. after that long flight. Are we in Liverpool already?"

"We sure are; the boys have gone to grab a taxi to take us to the docks. We don't have much time to spare though."

The Island beckons

The 'Lady of Mann' continued her stately passage down the Mersey past the Liver Buildings and Gladstone Dock and on rounding the Crosby Light vessel made a westerly turn to face whatever elements the Irish Sea cared to throw at her. Passenger accommodation was at a premium, while all deck and hold space was taken by seemingly hundreds of motorcycles, ranging from current models to veterans, all on their annual pilgrimage to the Isle of Man.

Following the evening meal some passengers retired to their cabins, while others adjourned to the saloon for post-prandial drinks. Rob was sensing a grudging acceptance from his companions as a worthy member rather than the new kid on the block, and as the evening progressed to a jolly conclusion he was intrigued by the actions of Dennis Mayne, the Midlands correspondent. For a time he had been accumulating a set of matches of varying lengths, when with a flourish he produced a cigarette pack with five matches protruding in line from its top.

"What's all this about Dennis?" Rob demanded. "Who draws the short straw?"

"Precisely old son," he replied, trying hard to disguise a devilish grin.

"And just what is the booby prize?" Rob enquired. "The loser buys the last round of drinks?"

"Stop being a wimp," Cyril King urged. "You're carrying on like some nervous bridegroom; just take a so-and-so match."

Rob preferred to be the last recipient, and at the final showdown his straw was clearly the shortest.

"Bad luck Rob," they all commiserated, each one clearly not upset with the outcome.

"It's all to do with the allocation of factory bikes for our personal use on the Island," King explained. 'We happen to know that four of them are the latest twins from Norton and Matchless, while the fifth is a 200cc James Captain, a fairly pathetic two stroke, and as holder of the short straw Rob, your transport during TT week will be 'Jimmy James'."

Rob gave a shrug of acceptance, privately wishing he had brought 'Clattering Kate' with him. That would certainly have elevated his standing in the present company.

With its atmosphere of old-world charm the 'Arragon' hotel was a pleasant enough base from where the team could conduct their TT duties. The Isle of Man was more than road racing circuit, it was a close-knit community, which for a number of weeks a year lives, eats and sleeps (if they're not too close to the circuit for early morning practice) motorcycling.

The team's immediate priority was to claim their machinery, and following breakfast they made their way to the aptly-titled 'TT Garage' situated at Douglas. Chris Beale, the proprietor was a former TT rider and was applying the finishing touches to his charges when they arrived. Two examples of the latest Norton Dominators plus two Matchless G9 twins were there for the enjoyment of the fortunate foursome, while to one side and looking rather forlorn was Rob's mount for TT week, 'Jimmy James', sporting a burgundy colour scheme and black pannier bags, which at least would be a great asset in his island perambulations.

Dennis Mayne and south coast correspondent Harvey Williams wasted no time in claiming the two 'Dommies', while photographer Bill Hanks and London-based Dave Nixon seemed content enough with their G9s. First item on the agenda was an exploratory lap or two of the circuit to check out any improvements or otherwise since last year's 1950 event. Rob had developed a mutual rapport with Bill who generously offered to 'puppy walk' him round this most challenging of circuits. Official early morning practice had just concluded, which seemed an opportune to set off, before the roads became clogged with commuter traffic.

Prior to embarking on his English odyssey Rod had endeavoured to absorb every aspect of the TT course, by way of photographs and circuit maps, however all research amounted to very little when faced with the actual reality. The first challenge was Bray Hill, just beyond the start, a steep descent that the stars negotiated absolutely flat-out, following which there was serious braking required to negotiate the Quarter Bridge right-hander. From there the course delivered a range of challenges, as one turn followed another in dizzying succession.

Even allowing for 'Jimmie's' modest performance Rob dared not lose concentration, as any deviation generally led to a painful meeting with an unyielding piece of real estate. Bill obligingly held back on any challenging sections, stopping occasionally for a photo shoot and concise description of that particular region. At the conclusion of their stop-start essay Bill estimated their time of 53 minutes represented a speed of 43 mph, and compared to the current lap record of 95 mph, Rob's admiration for these titans heightened a hundredfold.

TT practice week was a full-on affair for riders and correspondents alike, in particular those 5am starts, which tested all participant's stamina to the utmost. By virtue of his Press pass Rob was in a position to obtain great photo images of individual riders as they wheeled their machines to the start line. His presence however raised objections from a local photographer who presumed to have a monopoly on these activities. The situation reached an impasse with the local man making vigorous protests to the race committee concerning the interloper. These were quickly dismissed by that august body, with the recommendation that the complainant adopt a more charitable attitude to a visiting journalist.

For the remainder of the TT period an uneasy truce prevailed, somewhat souring what should have been a harmonious situation? Meantime, having gained a more intimate knowledge of the course, Rob was able to obtain graphic action shots at the more remote sections. For Monday's Junior TT he positioned himself at Hillberry, a spectacular right-hander, negotiated by the stars at 100mph. From there he was able to traverse open country on foot to include Cronk-ny-Mona and Signpost Corner in his photo shoot.

The culmination of a marvellous week of racing was Friday's Senior TT, which featured Junior TT winner Geoff Duke, then at the peak of a dazzling career as

No.1 Norton works rider. Rob's Senior tenure began at the start/finish line, from where he moved to the Bray Hill section and eventually to Quarter Bridge, one of the most deceptive of corners.

Duke did not disappoint his multitude of fans, leading throughout in a scintillating display of riding. For his part, Rob was well satisfied with his Isle of Man experience, to witness the world's best coming to terms with that most challenging of circuits.

Continental Capers.

Back at Bowling Green Lane, Rob had barely recovered his equilibrium from that Isle of Man experience when he was summoned to attend his next assignment, the Belgian Grand Prix. Located at Spa/Francorchamps in the verdant expanses of the Ardennes Forest this classic event had its beginnings in 1922, and offered every challenge from full-bore sweeping bends, demanding up-hill climbs to two first gear corners. Any aberration on the rider's part could see him in a perilous situation; like the Isle of Man, Spa demanded intense concentration.

Photographer Bill Hanks was also assigned to attend the Belgian, and being a keen sidecar exponent he was able to commandeer a lusty Vincent Rapide which had recently been the subject of a road test. Rob's three-wheel experience amounted to zero and he was quite content to be whisked along the continental highways in the relative comfort of a sporty Steib sidecar.

Official practice, which included a sidecar category, took place over two sweltering days, which was not uncommon in Spa in early summer. An overnight downpour on race eve was a welcoming feature, thus eliminating the prospect of melting tar. Race day was cool, cloudy and windless, offering perfect conditions for riders and charioteers. Once again Geoff Duke was in superlative form, winning the Junior race almost as he pleased, however the Senior GP was a much sterner affair, with Duke facing a team of four cylinder Gileras. Their far-superior top speed was offset to a degree by indifferent handling and the sheer brilliance of an in-form Geoff Duke.

Rob had positioned himself at the Eau Rouge bridge, a left-hander following the downhill start/finish straight and saw at firsthand what was remembered as the race of the season. Practice times earned Duke a place on the front row of the grid, alongside two of the Gileras and at flag-fall Duke made one of his meteoric starts. Was there ever a more thrilling spectacle as the vision of Duke hurling his

Norton through the left-hander and into the lead? Duke rode a copy book race, producing a ten/tenths effort, with both wheels drifting on those full, bore bends, the 'Champ' was in his element. This was also a red-letter day for Norton fans, with Duke delivering a Junior/Senior double and sidecar maestro Eric Oliver claiming his third successive Belgian GP.

From Spa the teams travelled north to the market town of Assen, which transforms into a fairground atmosphere with its influx of thousands of spectators to witness the Dutch TT. This event had its beginnings in 1927, and unlike the Isle of Man and Spa was completely flat, ten miles in length and with a mix of high and low speed corners. The Junior TT was a drama-filled event, with the unexpected retirement of Geoff Duke. Grabbing an early lead, but deceived by a strong tail wind he entered the left-hander past the start/finish straight far too quickly, laid the Norton over to an incredible angle and 'pranged' well and truly.

Fortunately no bones were broken and suitably patched up for the Senior TT, once again he held off the Gilera and Guzzi challenge to record another memorable double. With the Dutch TT successfully concluded, the 1951 season had reached the half-way point, with the French, Ulster and Italian Grands Prix yet to be contested. However for Rob and Bill Hanks, they would not be part of it. Having been summoned back to Temple Press; King and Williams would be covering those remaining events.

Brendan's Retreat

Rob had made a tentative flight booking for late August, leaving him two weeks to wind up his English connections. As a result of a comment to Bill Hanks about his own unique collection, the prospect of obtaining a replacement magneto for 'Bessie' was canvassed.

"There is a possibility Rob," Bill suggested. "A reclusive character on the south coast named Brendan Behan is alleged to have just about everything related to vintage Nortons squirreled away in his shed. I've got an address somewhere."

Bill produced a map of the United Kingdom, concentrating on an area around Hastings. "Here we are then," he announced, indicating a hamlet to the west of Hastings. "Unfortunately we don't have a phone number, so you'll just have to take pot luck and ride down there. I'm sure the editor will loan you my Matchless from the TT."

Rob's south coast odyssey began pleasantly enough, cruising down to Hastings in style aboard the new G9, and on nearing his destination a local shower was heavy enough for him to take shelter under a convenient elm tree. With the rain showing no sign of easing he retrieved a waterproof cover from a saddle bag and settled down under the tree to possibly catch up on some much-needed sleep.

At first unaware of the time factor he roused himself to realise that the rain had stopped and also that he had company. Nearby a lone motorcyclist was in the process of making adjustments to his machine. But what a unique bike this was! Rob was able to identify it from old photographs as a four-cylinder Belgian FN of around 1914 vintage; a rare enough bird in those days, but to chance upon one, some forty years later was simply staggering.

Bursting with excitement, Rob made his way to the scene, increasingly curious now about its rider. He was a rather a slim individual, immaculately dressed in a three-piece suit, cloth cap and highly polished shoes; hardly appropriate for an active motorcyclist in Rob's opinion.

His bike seemed to complement the rider's presentation, giving the appearance of being brand new, with that unmistakeable aroma of fresh paint and glossy leather. The rider looked up unsmilingly at Rob's arrival, almost impatiently at this invasion of his privacy. Rob presumed he was on his way to a rally of historic machines and ventured a comment.

"You've done a marvellous job restoring the old FN. Are you off to a rally somewhere?"

The man did not reply, instead he delivered a scornful look, which seemed to imply that the question was quite ridiculous. Rather deflated, Rob was still determined to contact the elusive Mr. Behan; perhaps this grumpy individual might enlighten him?

"I'm down this way to contact Brendan Behan," Rob enquired. "Could you possibly direct me?"

Once again the man remained infuriatingly mute, but at least gave a grudging nod of approval. Somewhat relieved Rob returned to the Matchless and booted it into life. He was not a moment too soon, for the mystery man was already on the move, pedalling furiously to gain momentum to start his engine. Off they went, with the FN setting a cracking pace for such a veteran. They negotiated various lanes and byways until a remote cottage came into view, and without slowing the FN rider pointed vigorously at the dwelling.

Rob pulled up at its front gate, complete with archway, climbing roses and a rustic signboard, which read 'Brendan's Byway'. Realising he would never have located the reclusive Brendan unassisted; Rob looked around to hopefully thank his guide. But to his surprise there was simply no sign of him; no cloud of dust, no burble of exhaust; nothing.

Totally mystified, Rob was at first unaware of a diminutive figure regarding him from behind the front gate.

"You managed to find the old place?" the man suggested.

Startled Rob turned to face him. "You must be Brendan; I'm Rob Barrett from Australia."

Smilingly Brendan opened the gate, issuing a greeting to step inside. Rob was faced with a puckish figure, barely five feet six tall and with such a beaming smile, Rob was instantly reminded of the affable Taffy Griffith."

"Welcome young man," said Brendan. "Step inside while I brew up a cup of tea and then you can tell me what brings an Aussie down here."

Brendan's kitchen was a pleasant enough situation, and the absence of any female company led Rob to presume he was either a bachelor or widower. After the usual small talk Rob was impatient to explain the actual reason for his visit. The fact that he was the owner of a pre-war Norton Inter was enough to broaden Brendan's grin, while mention of the Brooklands Special created absolute joy.

"I do have a problem with the B/S though," Rob pointed out. "I was conducting some speed trials back home, when the magneto casing disintegrated and now I'm battling to find a replacement."

"I may be able to assist you there, young Rob," Brendan suggested."But I have the feeling that something else is troubling you. Do you feel like unburdening yourself?"

Rob would have preferred to let the FN episode be forgotten, Brendan however became more insistent in his enquiries. Lucidly as possible Rob described the incident, at the same time feeling rather embarrassed.

"And this mystery man?" Brendan ventured. "Was he well dressed; you know; three-piece suit, cloth cap and polished shoes?"

"He was in fact," Rob replied. "I was beginning to believe it was one of those paranormal experiences."

"Perhaps it was young Rob," Brendan suggested. "That mystery man was 'Our Frank', making one of his appearances."

"Our Frank?" Rob echoed. "It's as though you knew him personally?"

Brendan's expression lapsed into one of reflective sadness.

"Frank Bateman was one of my best friends. This was around 1914, just before the Great War, and we were the local tear-aways on our new motorbikes; Frank on his beautiful FN and me on the latest Norton."

The mention of 'Norton' was enough to stimulate Rob. "What model was it Brendan?"

"She was a B/S like yours, but with lights and mudguards, and pretty crude compared to the FN, with its four cylinders and shaft drive."

"And what happened to our Frank?" Rob persisted.

At first Brendan was reluctant to continue his narrative, but with some cajoling from his visitor he laid bare the details.

"We were having a real dust-up one day, with Frank in the lead, as he generally was, when all of a sudden his bike goes into an almighty speed wobble, which wasn't unusual in those days."

"We call them 'tank-slappers' these days," Rob informed him.

Brendan smiled briefly at the term, and then continued.

"It was then that Frank came an absolute cropper. When I got to him he was stone dead, although there wasn't a mark on him. The local doc. reckoned Frank had suffered a heart attack."

"A nasty experience?" Rob agreed. "What happened to the bike afterwards?"

"After the funeral Frank's family put it into storage and it just sat there for years, until one day it just wasn't there. We presume some low-life stole it."

"Perhaps it was our Frank?" Rob suggested rather facetiously.

"We'll never find out Rob, but what I do know is that these visitations only occur in early August "Rob glanced at the wall calendar, which read August 4, the very date World War one was declared by Britain. Brendan made a significant comment.

"Now that you've had this experience I recommend that it never leaves this room in the future."

"Fair enough Brendan," Rob agreed. "Let's put it all behind us."

"There is another matter young Rob. You just finish your tea and I'll be back shortly." As good as his word, Brendan returned from his mission, bearing a compact cardboard box, and also the broadest of smiles.

"What do we have here?" Rob enquired, endeavouring to contain his curiosity. "Be my guest young Rob," Brendan replied, passing the mystery item across the table.

The carton was surprisingly heavy for such a modest size, and after a struggle to open it, he was faced with an object, still wrapped in its protective grease-proof paper.

"Go ahead lad," Brendan urged. "Let's see you unwrap it."

Breathlessly Rob complied, bursting with excitement. Almost in disbelief he was holding a well-remembered piece of history; a genuine Scintilla magneto.

"Brand new," Brendan assured him. "And it will fit your Brooklands Special like a glove. Let's call it a gift from Brendan."

For the moment Rob was stuck for words.

"Gosh Brendan, I don't know what to say."

"Was it worth travelling 12000 miles to get your hands on one?" Brendan suggested.

"It was Brendan. It surely was."

Ends

Author

Born in Sydney Australia, Murray McLeod is an accredited artist and author, with a number of self-published titles to his credit, dealing with vintage aviation history and motorcycle racing.

Murray was a keen motorcyclist throughout his teens and into later years. 'Tales of the Unexpected' explores paranormal events within the motorcycle realm and although entirely fictional is delivered in a manner which a motorcycle enthusiast will appreciate. Murray also welcomes commissions from interested readers for that special art work.

Artist/Author website

www.mcleodart.com.au

www.ingramcontent.com/pod-product-compliance
Lightning Source LLC
Chambersburg PA
CBHW060005230526
45472CB00008B/1950